A MEDICAL LECTURE

on the

Science and Philosophy

of

REPRODUCTION

By An Old Physician

FOR MEN ONLY

"Worth Its Weight in Gold

JAMES M. GRAY

First Printing 1890
Recopyrighted 1972, Thelma Mae Brackin Gray
Reprint Edition 1989
Recopyright Printing 2010

No portion of this material may be copied, printed, reprinted or reproduced in any form without prior written approval of the copyright owner, her assigns or her legal descendants. To do so will constitute copyright infringement and legal action.

ISBN 978-1-312-15395-0

Contact:
jamesmgray@bigpond.com
Brackinray Publications
11 Corndale Street
Loganholme, Queensland
Australia, 4129

Preface

This manual, first produced in 1890, portrays all the morals and scientfic thinking of medical doctors and scientific minds of the period. It is both comical as well as educational and contains information, much which if followed by men of this generation, would lead to a much more satisfying and worthwhile relationship between them and their respective partners. It will be enjoyabe as well as benificial to any who read it.

The lecture was not given to entertain the guests of the doctor, but to aid them in a better relationships with their wives and partners, and to instruct them with a means of enjoying a better and more fulfilling sexual experience; therby creating better family experiences.

Read & learn of the morals of a former generation and perhaps try the instructions of a hundred years ago.

Gentlemen: I promise to address you on this occasion upon a subject, deep in thought, which involves grand physiological questions, which may culminate in the deepest interest to your present and future welfare individually, and also in all of the relationships of life professionally, and especially to those who occupy the responsible positions as heads of families. It is not my purpose to arouse your indelicate thoughts, or to fire up your imaginations in reference to those animal passions, which not infrequently, are the out-croppings of hearts prone to evil, while I shall in chaste language, as far as possible, call things by their proper names. I shall at the same time, endeavor to enlighten your minds upon facts based upon the principles of science and morals, which I trust will result in the elevation of the purity of your hearts, and the promotion of your interest in all of the various relationships of life. The subject upon which I now propose to address you upon this occasion, is Reproduction. A distinguished professor in one of his lectures in defining Copulation, said, "It is a physiological act, it is the precedent of all other physiological acts, and assuredly is worthy of prominence before all others performed by earth born creatures from the lowest to the highest."

The history of Copulation dates from the very earliest period in the history of man. It is said that Adam, when he detected the *difference* between himself and Eve, he split the *difference*, and *'raised Cain'*, and that Mrs. Eve, as she smoothed her hair and put down the fig leaf said, "it is good;" that Adam, as he buttoned up his fig leaf with acorns, said "he would do it again when he got Abel."

The most widely known traditions of old, is that connected with the siege of Troy; there is no one who has not read of the theft of the beautiful bride of an ancient King, who was then an old King, and he was lacking in the qualities of a perfect man, in the fact that he was *too ripe;* in other words he was not in a bodily condition to perform the duties of a husband in a marital relation. So when Paris came he readily stole Helen from the old King's arms. In ancient times women were unfaithful, because of the badly regulated domestic relations, as they have been in all ages. Even, in the present history of the world, infidelity in women not infrequently, is the out-croppings of unfortunate marriages; old decrepit men with the frosts of many winters upon their locks become rampant for the beautiful and the gay, their brains are fired up with those imaginary passions which always characterizes old men on such occasions, and they at once seek some fair damsel in her teens and enter upon the wars of Venice, and when the time comes for the battle they realize that on "dress parade" they do well, but when attempting to storm the fort their guns will not go off, and they fall martyrs to contending forces, and almost in every instance, where the "flowers of May are united with the frosts of December," as in these instances the boideal of their affections becomes disloyal, and seek other fields for gratification of those pleasures which they fail to obtain from those aged veterans who have become deficient in their marital relatives. "The heart that is sick, must be fed," and they never fail under such circumstances to obtain relief from other sources.

Rarely, indeed, does a woman go astray, that the husband or father is not directly to blame for it. They do not want to do ill, these luckless ones, for they are luckless at the very best, they only want to be cared for. A woman fitly grows doubly good, the reverse of the proposition, however, is not true necessarily, and it is just as well for men to remember, that women have a little human nature in them, and that, men equally with women, safely leans to virtues side, most certainly to the virtue of manly love and care, an honorable, domestic respect and duty, above all, personal supervision and sympathy. Perhaps in no station on the inhabitable globe are to be found among women, more virtue and purity of character than our American daughters possess; and in all of the relationship of life are preeminently superior to those who claim the priority. That some, in hours of temptation may fall, none will deny; but these are obligations of momentous import resting upon every man who has a mother, wife, daughters or sisters to stand out in the defense of nature, and sustain those who may have fallen, and raise them, if possible, to higher walks of life, and lead them on to noble instincts which characterizes those which are not infrequently recognized as Angels of mercy to the sorrowing and the distressed.

You may ask me, what has my talk thus far had, except in a very indirect way, to do with the subject of Copulation? I may answer by saying that it has identified with it the equality of love, and *vice versa*. The eunuch is cold, crafty, and deceitful, a thief, a liar, a murderer---anything that is absolutely devoid of the first instinctive idea of love. We have installed into us, from early life, that sensuality is sensuality and love is love. Poets tell us that love is a principle, that love is purely intellectual passion, and that sensuality is as distinct from it as the stars are from the sun. The sun warms and invigorates, and the light, flashing on every part of the globe, reflects the penetrating powers of love, while the stars, dim with no radiating heat, leaves sensuality, involved in gloom and covered with the damps of death. There is unmistakably a distinct and wide difference between the finite power of the man and the woman, as in the quality of love. It is said they are antagonistic, and they, like extremes, meet and come together to produce a harmony, that makes, creates, the living creature. There is said to be a disposition on the part of the man to cultivate the sensual and by the neglect of the higher and nobler, and better half of life, he debases manhood. Undoubtedly there is truth in this, undoubtedly there is a stronger tendency towards the sensual side than the higher and purer in man, and by this tendency and its cultivation, man, unfortunately, falls below those qualities with which nature endowed him, and brings himself but a few degrees higher than the brute. Perhaps in no way does man so utterly destroy himself as in that awful act of masturbation. Thousands of young men, throughout the entire world, are the woeful subjects of this practice. This departure from the designs of Nature, can be traced to a very early period in the history of the world. The first violator of this law was an Israelite, under the Jewish dispensation, by the name of Onan. The Lord directed him to cohabit with a certain woman. I suppose the Lord saw that Onan was in need of relief, but in spite of this direct command he turned a deaf ear to the command of Jehovah, went out, and by an ingenious manipulation of his hands, "split on the ground."

Now gentlemen, I am prepared to state, in the most explicit terms, that the effect of this horrible practice has been the downfall of many of the brightest young men of the country. What are its effects upon the human organization? It utterly paralyzes the whole nervous system, by acting upon the base of the brain, and its connections with the spinal marrow, destroying the digestion, and utterly rendering these unfortunate subjects unfit for business, and totally subverting all of the social relationships of life; while many of them, not infrequently, become the inmates of insane asylums.

The basis of my lecture is this: that a large element in domestic and social happiness, is the ability on the part of both male and female, to perform properly the marital act. The cause of my lecture is largely to the fact that it is very common for women to report to their doctor, or family adviser, that they are lacking in the one quality, namely, the enjoyment of the sexual act. It is a matter of every day occurrence for men to report that they are lacking in the ability to perform the sexual act; and it is a well known fact that a want of adaption and harmony between the sexual organs of male and female, not infrequently prohibits the performance of the act entirely.

There is a tendency on the part of the

male, to frequent erections and premature discharge, a tendency which is very common with the male, when the erection is accomplished so as to prematurely complete the act before there is any possible chance to gratify the female with whom he may be in contact. There is undoubtedly a remarkable failure on the part of a very considerable percentage of the women to ever enjoy sexual contact. I know women who have three or four children, who declare they have no idea what it is to be sexually excited. I have had women to consult me often, and say: "What is it, and how is it, that this enjoyment of which I have heard so much, pass out so poorly." They say that, "while my husband gets his pleasure, it leaves me ungratified and thoroughly disgusted."

Young men consult me about the following: "Doctor, I want to get married, I am in love, and I am engaged, but I put it off, because when I was a boy I masturbated a good deal, and I am afraid I have destroyed my manhood. I have stopped a good while, but I have about three emissions a month, and I am afraid I am hopelessly affected." Such men are generally answered as follows: "You say you are a clerk? Very well. As you come out upon the sidewalks you mingle in a crowd of brokers, and men of all trades, professions and crafts, now if you will bring me one of these men who has *never* masturbated, and who will *swear he never has,* I'll swear in my turn, he is either a liar or a fool, and my advise to him would be as a part of worldly education to go and masturbate once to see how it goes." I say to him, "you frequently meet troops of rosy, healthy, happy school children tripping homeward, laughing, frollicking, rolling and tumbling about in the snow, and perfect pictures of health and beauty. These, in some instances, are the children of masturbators. It is so nearly universal, that it is safe for the philosopher to say that it is universally practiced by the male portion of the human family. It is a popular notion, among men, that because they did so when they were boys that girls are no less guilty than they. It is grand to contemplate the purity of female character upon this subject, they soar far above such grovelling passions, and spurn from their minds the first intimation of feeling so contridicted in pure womanhood. It has been part of my policy in the practice of medicine, to have no hesitation in asking any reasonable question of a patient, and I have not a few times put this question. I have asked it of prostitutes, I have asked it of girls who were not exactly that far davanced in lewed life; I have asked women respectable in every sense, and I must say that it is the exception among the girls.

I believe that much of the diminished pleasure in sexual congress, may be directly referred to masturbation, and I believe, and must insist upon the principle, that some knowledge is necessary to accomplish the marital act perfectly and completely, and secure its legotimate ends of entire gratification and reproduction. Masturbation, I regard as a debacing act, an unmanning act, a disgusting act in all of its features, and it presents not one characteristic that place it above the act of the brute.

I know a young man about 18 years of age, smooth face, blue eyed, fair skin, who says he cannot work, that he has ruined himself by self abuse, and his comrades in the shop here raised up a purse for him to pay his expences because of his inability to make a living. He has not abused himself for two years, and he was not capable of discharging seman more than three years previous to that. The generative organs are similar to that machine which inventors have been cracking their brains to construct for years past, the machine of perpetual motion, and when a man once starts his machinery it is inclined not to stop, until he passes beyond the time of verility. A man, then, once starting the process in boyhood, beginning to reason about it, is prevailed upon to stop it by his manly sense, and is soon after disgusted and chagrined to wake up in the morning, to find that he has spilled a United States Senator on his sheet. He starts out in the morning disabled with pain in the loins, and headache from the unnatural discharge; and in reference to this I may say that any discharge irregular or not in accordance with nature or that is not in perfect harmony with the mental, moral and emotional state which was intended to accompany it, is far more exhausting than that which is in such accordance. The gay Lothario with his mistress in seclusion perhaps commits far greater excesses than men do when legitimately engaged, yet he leaves in the morning as gay as a lark.

Some years since when discussing the subject with a medical man, he remarked: "The sensation is delightful, but the attitude is ridiculous, and the recoil doleful." Yes, the reason, I said is this, you have an engagement with your feet on the tender at

8 o'clock, and precisely at that time you are there with your feet to the fire, perhaps watch in hand, waiting for the creature, wondering if she will come, when at the minute precisely, in she bounced with the exclamation: "Well, Doc I'm here you bet !" You hold a sceance with her, pay her wages, and she leaves, and you are left alone to your reflections. Now you are an intelligent creature, you are a reflective man and thoughts of this kind pop through your miond. "That's a dirty bitch; Wish that I had never seen her. That'll be the last time." You are in that extremely virtuous frame of mind produced by the sexual act. You know the only rational explination that can be given for the answer of Mr. Joseph to Mrs. Potiphar is, that Joseph had just had some.

The point I wish to make in this connection is, that a man under such circumstances, feels justly debased and disgusted. I hold, gentlemen, that a large majority of the social ills to-day are directly traceable to their ignorance, inability or neglect on the part of one or both sides of the family in this matter of sexual congress, and that a man owes largely his incompetency to this early practice, and if everybody could have it impressed upon him that he is destined to lose untold pleasures by his unnatural practices, these ills would be to a large degree obviated. He becomes accustomed to such a morbid excitation and such a poor substitute for the natural stimulant to his powers, that when the physiological conditions are brought to bear upon his organ, it cannot stand the pressure. No matter whether the hand used to accomplish such base desires has held the plow, or been subjected to no more ponderous weight than turning the leaves of a book. It is compared to the proper stimulus, rough, cold altogether a very poor substitute; and the organ trained to such coarse contact, when it meets the proper society, prematurely gives up and hangs its head in shame.

If there could be brought together a race of men and women alike educated, and in whom sexual excitement had never occured, I believe there would be a greater conformity in the sexual gratifaction than there is. As it is, it is the male portion who should be informed how to perform properly. Men come not infrequently and say: "Doctor, I am engaged to be married but I am impotent, I don't dare to marry; I am satisfied if I should I would bring both myself and wife to unhappiness because I have lost my manhood." How do you know you have lost your manhood, don't you ever have an erection? "Oh! yes, frequently; why even while I sit beside the woman I adore, much as I love her, I cannot help that. But I have tried myself. Three or four times I have gone down on the 'chute' and given my money to some daisy, and I can't get my organ erect. Now just fancy me in that condition as a married man! I'd commit suicide.

Such an occurance as this conversation comes to the man who has the work of drawing out the confidence of young men, in an experience of forty-five years. Now whom sexual excitement had never occured, I believe there would be a greater conformity in the sexual gratifaction than there is. As it is, it is the male portion who should be informed how to perform properly. Men come not infrequently and say: "Doctor, I am engaged to be married but I am impotent, I don't dare to marry; I am satisfied if I should I would bring both myself and wife to unhappiness because I have lost my manhood." How do you know you have lost your manhood, don't you ever have an erection? "Oh ! yes, frequently; why even while I sit beside the woman I adore, much as I love her, I cannot help that. But I have tried myself. Three or four times I have gone down on the 'chute' and given my money to some daisy, and I can't get my organ erect. Now just fancy me in that condition as a married man! I'd commit suicide.

Such an occurance as this conversation comes to the man who has the work of drawing out the confidence of young men, in an experience of fourty-five years. Now and emotions of the human heart.

So this young man, in this condition, I gave this prescription, which is not justified by moral sense, but which is clinically unsurpassed from the fact that it is generally quite sufficient without being filled. Now if you are hinging your whole future upon your ability to copulate with one of these prostitutes, the only way you can accomplish that is to make a night of it. Go and play cards, and drink until your moral sense is so blouted that these thoughts and pictures I have presented can have no effect upon your mind and there will be no trouble in coming to time. It has been said that alcohol creates the desire and destroys the capacity. You will find, if you take the trouble to examine the penis, hard from alcohol, no less than the man hard in its body and soft in its head.

Now, I have fore-shadowed that the cause of failure in copulation is the abuse in boyhood and manhood, of the organ by which it is accomplished, and I have maintained the point that the vast majority of the girls do not masturbate, that it is à rare thing for them to do so. Now how do they masturbate? The candle theory is ridiculous, and a damning libel on the purity and chastity of our women. I had a married woman come to my office, who by her lacivious actions and lewd talk inclined me to the belief that she was borded on *Nympyomania*. She had played her husband out. I asked her what she did in such matters, presuming that she was unselfish in the liberality with which she bestowed her pleasures upon others. She replied that frequently she put herself off. I continued inquisitive, and asked to know how. She voluntered to illustrate, and lying down on the lounge, she simply rubbed her hand and fingers over the external genitals. There was simply contact; no entrance, whatever. She was a *Nyphomaniae,* as has been demonstrated by her confinement to an asylum. Now, gentlemen, I believe these points have no little bearing upon the proper performance of the act, as well as your domestic happines and pleasure.

Some years ago a farmer came to me, very much excited, who had a wife who was a good wife, and in the course of event of his farm life, a red-headed, blear-eyed Irishman presented himself to him, and being in need of help, he employed him to work. The upshot of the transaction was that one day the Irishman and the woman disappeared. After a long search the farmer found them living together. He entreated her to return with him but all in vain. There again we have one of the ten thousand illustrations of domestic infelicity. The man was not impotent, because he had two children. He was incapable of retaining his wife's love because of his ignorance in this matter. He was an ignorant man, ignorant of what constitutes pleasure to the female, and ignorant probably of the fact that he never excited other than a feeling of disgust in his wife.

It may be said, in this connection, that there appears to be an antagonism, and is wonderful it should be so---between the means of producing pleasure for both parties at once, which in other words means that the act which produces the most pleasure to the male, produces the least pleasure in the female, that if the male performs that act after the fashion of masturbation, he will gratify himself, but rarely succeeds in creating any pleasure in the female. There are many men who would go across the river on the floating cakes of ice, taking the chance of drowning to do an act of kindness to a suffering woman; there are men who would take almost any risk to gratify their lady love, and yet they have never succeeded in doing so, because of ignorance and incapicity. I have lately received a communication from a woman, the wife of a medical man, saying she has no sexual gratification, and asking me if I can advise any medicine by mail?

The woman who finds herself incapable of enjoying the sexual act frequently becomes morose and unhappy; and the man who finds he cannot gratify her, will consult the doctor to find out what the trouble is. The man can go elsewhere, and if the woman fails to get the gratification she is led to expect, it is but the part of human nature that she should look elsewhere also. It is exceedingly possible that the husband will look elsewhere for a woman who can act second part to the tune he wishes to play.

I am coming rapidly to the front as to how the male should perform the sexual act. One of these afternoon sewing societies met on a certain occasion. A number of the ladies got together after lunch; they got to discussing family relations, and presently the matter of sexual congress came up, and said one of the ladies: "I don't like it." Another said: "I think it is the nastiest thing in the world. I put clean sheets of the bed every Saturday night, and William soils them all up at once." Another broke in with, "If there is anything in this world that makes me wish I had lived to be an old maid, it is to have John come home half full at 11 o'clock at night and say, "Lay over old woman." And thus it went on, the society and all of them, crying down the sexual act as nasty. Well, under such circumstances, it is.

But amongst these women there was one demure little black eyed, black haired woman, who had not laid down her work. She was making an elegant night shirt for her husband, and she had taken no part in the conversation. Presently, as with one accord they turned upon her and said: "Mollie, what's your opinion of this subject? Now we have always noticed that there never was a happier couple than you and your husband, Charlie. You seem to be as much lovers as ever, and you have a little boy; so now Mollie, What is it? Speak out, Tell us your opinion of this thing? She said: "Ladies, excuse me if you please, I don't consider such topics fit for general conversation, and I don't believe in converting our society meetings into such discussions. But they insisted, and pulled away the night shirt, needle and thread, and said: "Now Mollie, we are bound to have your opinion. You're in for it," until she finally said, "Well if you must have my openion, I'll tell it to you. "Quite frequently of nights when Charlie comes home from the office tired, he comes to bed, and after kissing me, puts his cheek against mine, and soon we are fast asleep, and no more until we wake up to romp with Little Charlie in the morning. But sometimes, when he comes he dosen't seem to be sleepy, and he puts his arm around my neck, and he runs his fingers through my hair, and says: "How beautiful your hair is, Mollie; it reminds me of the days when we were courting, when I used to love to run my hands through it. And later his hands will stray down, and unloosing my gown until my berasts are exposed, and he says, "why, Mollie dear, these are just as beautiful as the day we were married; they are not the least impaired by the sucking of our darling child. Why, this little red nipple sets up here like a strawberry on a saucer. And finally, when his hand strays down upon my person, and he says: "Does the little birdie want a worm? Oh! My!"

Now, gentlemen, you can bet all you are worth against a ten cent novel, that there is a woman who enjoys sexual intercourse; and there is a man that knows how to treat a woman and retain a wife's affections. He dosen't say, "Lay over here old woman!" He is not one who only seeks his own pleasure, first the pleasure of his wife, and then his own.

There are thousands of men who regard women merely as animals for the gratification of those base and grovelling passions which dwell only in the hearts of men who sympathises for the beautiful and the lovely are but second considerations when brought in contact with their animal nature. Every man should regard his wife as an Angel in human form; as the sensitive plant to be touched only by the finger of love. Even in those sexual enjoyments which is connected to the married life there are limits which should not be gone beyond. No man who regards the health of his wife, should trespass upon her modesty more than once or twice a week; when beyond this, it is an infrinement upon her marital rights, tending to the prostration of her nervous system and general ill health throughout life.

Not long since a lady came into the office of a prominent physician of Atlanta, in an excited state of mind, even sheding tears. The doctor inquired of her the cause of her distress. She declared to him that she was ruined, her health was gone; and when asked for the cause, replied that her husband had destroyed her health by his frequent intercourse with her. The physician asked her how many times during the night. She modestly replied: "Five or six times, and that she had actually given him three dollars to go somewhere else for her relief." Such men remind me very much if a certain animal who has long ears, and but a small brain caliber, and I know of but one remedy for such indescretion in men of this character, but *castration*.

A large majority of females in this

country who are the subjects of diversified diseases, are traceable to the discretions of thoughtless husbands. If the evil was to end here, it might be some alleviation from censure; but it is a physiological fact, that children begotten under such circumstances are most generally frail, and their nervous systems greatly unstrung, slow in growth and most generally find premature graves. I might add, as a great scientific problem in relation to children, by way of securing them to sound constitutions an insurance of longevity, a fact which can only be obtained through the grand teachings inculcated through Phrenology. Nature utilizes variety of intemperaments in producing healthy and vigorous offspring. The union of two nervous temperaments most generally results in the production of frail and unhealthy children; while the union of nervous and sanguine, or billious and nervous are productive of the opposite results of the other. The plegmatic united with the highly, sensitive in a great degree lessons the sexual enjoyment. In order to enjoy the highest results of this physiological law in married life, the husband and wife should be of different temperaments.

In marriage, it should be the purpose of all parties to seek light from the scientific principles inculcated in Phrenology, and no young man or woman should risk their destiny upon the uncertain sea of matrimony without being well instructed upon the unerring guide of Phrenological law.

The next important question which demands our investigation, is how shall copulation be performed to excite pleasure in the female? I hold no man should enter a female if he finds the parts dry and chippy, as the palm of the hand, because this is sufficient evidence that the female is not in a condition of body or frame of mind to enjoy the sexual act. She will either be disgusted if not made to actually suffer by it. Under these circumstances men are apt to flatter themselves, that they are too large for women. Such an erroneous openion is but completed. If an attempt is made to overcome a female against her will, the discharge will almost invariably occur at this stage, or even before the penis comes in contact with the parts at all.

There was never a woman raped yet that was in earshot of man but was half willing. No woman in her senses can be entered without her consent, but with her ability to scream and her powers to bring her thighs together, unless there is some willingness on her part, he be ever so strong and large, and she ever so small and delicate, unless she loses consciousness, he cannot effect an entrance. And why not? The man who enters into this hard task will find that his penis will drop inevidently, because the excitement of all his desires drains the tide of blood necessary to sustain an erection away from the penis, and it will irrevocably fall.

Then when a man attempts to enter a woman, and he finds he is about to complete the act, and then he feels that if he does new dress she got down town to-day? You will be surprised, when I tell you that men frequently tell me that by the exercise of judgement, they may prolong and perform, the act for half an hour.

That, then, is the first thing to accomplish, to get the penis accustomed to being placed in the moist, hot parts. Then, the organ being carried in its entire length, if the male plays it in and out as if in a sheath, he may complete the act so far as he is concerned, without having given the woman a particle of pleasure. At the upper part of the internal parts is placed the sensitive clittoris, and during the proper performance of coipulation this clittoris undergoes erection and becomes larger and larger as the womans pleasures are increasing, until it stands out erect, and receives the brunt of the excitement in the act. A man remembering this will adapt himself to titillate that clittoris, while he knows it not will play up and down and will excite only the duller portion of the generative tract. That man who rides high push it once more, a little further, it will be all over with him, he should shut his teeth firmly, and in spite of everything withdraw and prevent the escape, and thus prevent the completion of the act. I direct those men who are in danger of prematurely completing the act, that they introduce the penis very closely. After the first few times, with a young wife, if the head of the penis can once be buttoned in before the discharge occurs, then the mans trouble is over. The trouble is, that in driving the member into the canal, the man will find that the hot most parts are rather too strong for him to stand.

A man who has been accustomed to drinking light wines is not able to navigate when he drinks brandy. So then, when a man finds himself on the verge of completion of the act, he should take a rest, stop and talk, there is no objection to his stopping right there, anmd talking to his wife. That will divert the mind and give them time to get better acquainted, as it were. He might ask her if she paid for that

Luther, I suppose that is allowed among religious circles. The man who has created the greatest disturbance, looked at from one standpoint and the greatest advance from another has said:

"Who loves not wine, women and song
Live's a fool all his life long."

The import of which is that, if a man copulates twice a week, he ought to live to be 104 years, but when beyond this, he developes all the characteristiocs of the brute, and entails upon his wife untold miseries the remainder of life. But Luther came late in life into the field of action of man against woman, and this doctrime of Martin's will not stand the test of ages.

The most interesting question is, however, how long a man or woman is able to copulate? I am almost afraid to venture an openion on that; but as an illustration, I will give you the experience of a venerable old preacher at a camp-meeting. He said: "Bretheren, I stand among you as an and goes slow, is the one that titillates the clittoris, and can withstand the degree of excitement almost indefinitely, and until the woman has been entirely gratified, before he accomplishes the act himself.

Bear in mind that you should do unto others as you would be done by; bear in mind the golden rule in this as in other relations; and this, the great living tie between man and woman may be indissolvely bound, and unite you through a long, happy and prosperous life. Many other points might be mentioned, but the cardinal point is, for the man not to push until the desire to complete the act passes off, and then he can resist it until it suits his pleasure and convenience, and the next point is to remember that it is not the length but the proper motion that titillates the clittoris, and the motion that is most gratifying to the female is least gratifying to the male, and thus enables him to complete the act at his pleasure.

The greatest of all reformers was Martin

ancient tree in the forest of the Lord, the frost of seventy winters have scattered my leaves to the four winds of heaven; the winds of seventy winters have bent and twisted my limbs, the ice of seventy winters has cracked and shivered and peeled the bark from my trunk; but thank God, my old root still stands."

Storms may howl and suns may set,
    I'm still a man for all of that,
Though often chilled by the freezing blast,
    But I'm sure to get there at the last.

Copyright 1972/2010
All Rights Reserved

www.ingramcontent.com/pod-product-compliance
Lightning Source LLC
Chambersburg PA
CBHW080851170526
45158CB00009B/2706